Oskar Becker on Modalities

Stefania Centrone & Pierluigi Minari

T0107837

Philosophische Hefte

Band 10

Herausgegeben von

Prof. Dr. Axel Gelfert
Prof. Dr. Thomas Gil

Oskar Becker on Modalities

Stefania Centrone
and Pierluigi Minari

Logos Verlag Berlin

λογος

Philosophische Hefte

Herausgegeben von

Prof. Dr. Axel Gelfert
Prof. Dr. Thomas Gil

Institut für Philosophie, Literatur-, Wissenschafts- und Technikgeschichte
Technische Universität Berlin

Bibliografische Information der Deutschen Nationalbibliothek

Die Deutsche Nationalbibliothek verzeichnet diese Publikation in der Deutschen Nationalbibliografie; detaillierte bibliografische Daten sind im Internet über http://dnb.d-nb.de abrufbar.

ISBN 978-3-8325-5031-8
ISSN 2567-1758

Logos Verlag Berlin GmbH
Comeniushof, Gubener Str. 47,
10243 Berlin

Tel.:   +49 (0)30 / 42 85 10 90
Fax: +49 (0)30 / 42 85 10 92

http://www.logos-verlag.de

Affiliations

Stefania Centrone

Institut für Philosophie, Literatur-, Wissenschafts- und Technikgeschichte
Theoretische Philosophie
TU Berlin
Straße des 17. Juni 135
10629 Berlin
Email: stefania.centrone@tu-berlin.de

Pierluigi Minari

Dipartimento di Lettere e Filosofia (DILEF)
Università degli Studi di Firenze
Via della Pergola 58-60
50121 Firenze
Email: pierluigi.minari@unifi.it

Contents

1 Introduction

This booklet aims to present, to contextualize and to evaluate the pioneering contributions to modal logic contained in OSKAR BECKER's essay *On the Logic of Modalities* (*Zur Logik der Modalitäten*) that appeared in 1930 on the *Yearbook for Philosophy and Phenomenological Research*.[1]

OSKAR BECKER (Leipzig 1889 – Bonn 1964) was a German philosopher, logician, mathematician and historian of mathematics. He is often remembered, together with MARTIN HEIDEGGER, for being one of the most important students of EDMUND HUSSERL (1859–1938). He was, together with MORITZ GEIGER (1880–1937), MARTIN HEIDEGGER (1889–1976), ALEXANDER PFÄNDER (1870–1941), ADOLF REINACH (1883–1917) and MAX SCHELER (1874–1928), one of the members of the editorial board of the *Yearbook*.

OSKAR BECKER got his PhD in mathematics in 1914 with a work[2] entitled *On the Decomposition of Polygons in non-intersecting triangles on the Basis of the Axioms of Connection and Order* (*Über die Zerlegung eines Polygons in exclusive Dreiecke auf Grund der ebenen Axiome der Verknüpfung und Anordnung*). In 1922 he wrote under HUSSERL's supervision his *Habilitationsschrift*, *On Investigations of the Phenomenological Foundation of Ge-*

ometry and their physical Application (*Beiträge zur phänomenologischen Begründung der Geometrie und ihrer physikalischen Anwendungen*).[3] In 1927 OSKAR BECKER published in the *Yearbook* his masterpiece *Mathematical Existence*,[4] where he uses the Husserlian phenomenology to clarify the process of counting. In 1952 — when the study of modal logic was already well beyond its pioneering era — BECKER came back to the subject publishing a monograph, *Investigations on the Modal Calculus* (*Untersuchungen über den Modalkalkül*), perhaps too old-fashioned for the time.[5]

* * *

The essay *On the Logic of Modalities* represents an attempt to treat modal logical issues with a phenomenological method. This enterprise appeared from the outset not to be easy at all, for logic and phenomenology are completely different disciplines. Depending on the way in which it constructs its formal systems, formal logic can be seen as *the theory of the correct inferences*, or alternatively, as the theory *of purely formal truths*, that is, as the theory of those truths that hold without any condition. Phenomenology, instead, deals with the description of lived experiences.

Indeed, we might better say that in his investigations BECKER pursued two loosely related goals.

The first one, more technical in character, was to find axiomatic conditions that reduced to the finite the number of logically non-equivalent combinations arising from the iterated application of the operators "not" and "it is impossible that (...)" in LEWIS's modal system, as we will explain in details below. The second one, more philosophically oriented and in a sense much more ambitious, was to treat the logic of modalities from a phenomenological perspective and to understand, from this perspective, the philosophical and logical-ontological problems underlying the, and posed by, Intuitionism.

On the Logic of Modalities consists of two parts, loosely related as the above mentioned corresponding goals are. Part I contains a general Introduction that shortly recalls the Aristotelian conception of modalities as well as HUGH MACCOLL's pioneering modal logical investigations in his *Symbolic Logic and its Applications*[6] of 1906. It then focuses on C. I. LEWIS's *Survey of Symbolic Logic*[7] of 1918. This latter contains the first presentation of the so-called "*Survey* system", known since 1932 as "modal system **S3**."[8]

De facto, **S3** is the actual object of the investigations in Part I of BECKER's essay. As pointed out by EMIL L. POST, the system LEWIS presents in 1918 *collapses* into classical logic. LEWIS corrects it in a paper entitled *Strict Implication: An Emendation*[9] and published in 1920, where the system effectively becomes the logic we nowadays know as "**S3**."[10] In his essay BECKER faithfully reports both that the original version of the "*Survey* system" proves the *collapse* of modalities, as well as LEWIS's amendment thereof. Incidentally, "*collapse of modalities*" is a customary expression in the modal-logical jargon. It means that a modal logical system proves that *necessity* and *truth* are one and the same, or equivalently (as it is the case in the "*Survey* system") that *impossibility* and *falsity* are one and the same. Obviously, such a system is *trivial* from a modal point of view.

BECKER's Introduction touches on the paradoxes of *material* and *strict implication* and sets out to establish a propositional modal logic that is *decidable* as the classical propositional logic:[11]

> The aim of the present essay has a strict relation to the investigations of MACCOLL as well as to those of LEWIS. The ultimate purpose of our investigations is to develop an elementary logical calculus that takes adequately into account the *modalities* of the sentence, namely in such a way *that the so-called elementary decision problem is solvable*, as in the ordinary propositional calculus.

Part **I**, *On the Rank Order and the Reduction of Logical Modalities* — on which this booklet will concentrate — is specifically devoted to the problems of ranking and iteration of modalities. BECKER sets out to modify **S3** by means of some additional axioms effecting the reduction of complex modalities to simple ones in order to obtain two new modal systems — he calls them "*the six modalities calculus*" (henceforth denoted here by **S3'**) and "*the ten modalities calculus*" (henceforth **S3''**) with the following properties:

(i) *the number of irreducible modalities is finite,*

(ii) *the positive* (and by consequence the negative) *modalities are arranged in a linear order with respect to logical strength.*

He believes that, since the "System of Strict Implication" has the conjunction, the negation and the impossibility as primitive logical constants, it is possible to generate within it infinitely many non equivalent nested modalities through iteration of the negation and the impossibility operators. Such modalities, as KURT GÖDEL (1906-1978) puts it in his *Review of* BECKER *1930*, "cannot even be linearly ordered according to their logical strength in the sense that, of any two affirming modalities, one will imply the other, and similarly for negating ones."[12]. Otherwise said, there are modalities that are incomparable in LEWIS's system.

That said, it is worth to be mentioned that OSKAR BECKER neither shows that the two systems he sets up (and others he tentatively introduces, as we will see later) really differ from one another, nor that his additional axioms cannot be derived from those of Lewis, nor either that in his own systems, with *six* and, respectively, with *ten* "irreducible" modalities, such modalities cannot be further reduced.[13]

Actually, nine years later, W. T. PARRY will show, in a paper entitled *Modalities in the Survey System of Strict Implication*[14], that, at a variance with what BECKER seems to believe, **S3** has a finite number of modalities. More precisely, PARRY shows, with the help of a number of suitable theses he is able to derive in the system, that it is possible to reduce all the complex modalities in **S3** to a finite number of irreducible modalities, viz. 42. He also shows that no further reduction is possible.

Part **II** of BECKER's essay explores, more or less independently from Part **I**, the connection between modal and intuitionistic logic both from a formal and from a phenomenological perspective. From a formal perspective, the particular interest of a (propositional) modal calculus with nested modalities that is *decidable* lies in the fact, so BECKER, that BROUWER's idea to set up a finite logic grounded on *evidence*, or – to put it with HUSSERL – on the *clarity of evidence* (*Klarheitsevidenz*) seems to be realizable only within the framework of a modal formal system.

Indeed, BECKER is the first logician and philosopher of mathematics to put forward the idea of a *modal interpretation* of intuitionistic logic, more precisely the idea of a possible sound and faithful translation of intuitionistic logic into modal logic. However, the first actual translation is to be found in a one-page celebrated and influential paper entitled *An interpretation of the intuitionistic propositional calculus* written in 1933 by KURT GÖDEL.[15] The basic idea of GÖDEL is similar to the one OSKAR BECKER outlines in *On the Logic of Modalities*.

BECKER suggests to add to classical logic the predicates "(...) is provable", "(...) is such, that its negation is provable" and "(...) is undecided". Such predicates should express BROUWER's primitive logical concepts.

Similarly, GÖDEL's idea is to add to the language of classical propositional logic the unary operator "it is provable that (...)", denoted by "B", and to an axiomatic calculus for propositional classical logic *three* axiom-schemas and *one* rule of inference. The axiom-schemas are the modal schemas K, T and 4 that characterize modal logics that are nowadays standard, the rule of inference is the necessitation rule that is contained in all *normal* modal systems.

We will introduce both the schemas and the rule of inference in detail later on.

Notice, incidentally, that both BECKER and GÖDEL seem to take the predicate "(...) is provable" and the operator "it is provable that (...)" as conveying the same piece of information. Actually, the predicate "(...) is provable" denotes the property of a proposition to be provable, while the operator "it is provable that (...)" takes a proposition as input and gives a different proposition as output. (Unfortunately, such practice of systematically neglecting the difference between predicate and operator is, even nowadays, quite widespread among logicians.)

GÖDEL writes:[16]

> One can interpret Heyting's propositional calculus by means of the notions of the ordinary propositional calculus and the notion "p is provable" (written "Bp"), if one adopts for that notion the following system $\mathfrak{S}$ of axioms:
>
> 1. $Bp \rightarrow p$
>   if it is provable that p, then it is true that p
>
> 2. $Bp \rightarrow ((B(p \rightarrow q) \rightarrow Bq)$
>   if it is provable that p and it is provable that p implies q, then it is provable that q

3. $\mathrm{B}p \to \mathrm{BB}p$

if it is provable that p, then it is provable that it is provable that p

In addition, [...] the new rule of inference is to be added

$$\frac{\mathrm{A}}{\mathrm{B}A}$$

From A, it is provable that A may be inferred

By substituting throughout the operator "B" ("it is provable that (...)") by the operator "□" ("it is necessary that (...)") one obtains one of the modal logical systems that are nowadays standard, namely LEWIS's system **S4**.

2 The Conditional, or *The Crows on the Roofs*

Since BECKER as well as MACCOLL and LEWIS all refer to the old controversy about the right interpretation of conditional sentences, let us briefly dwell on it.

Such a controversy traces back to the Megarians and the Stoics. As JÓZEF MARIA BOCHEŃSKI puts it in his *A History of Formal Logic*:[17]

> The definition of implication was a matter much debated among the Megarians and Stoics: All dialecticians say that a connected (proposition) is sound, when its consequent follows from its antecedent — but they dispute about when and how it follows, and propound rival criteria.
>
> Even so Callimachus, librarian at Alexandria in the 2nd century B.C., said: 'the very crows on the roofs croak about which implications are sound'.

In ancient times the quarrel was, above all, between a truth-functional and a modal interpretation of the conditional. PHILO (OF MEGARA) said that an implication is true when it is not the case that it begins with the true and ends with the false.[18] This conception of the conditional was later adopted by

GOTTLOB FREGE (1848-1925) and by the American logician and founder of American Pragmatism CHARLES SANDERS PEIRCE (1839-1914).

In his *Gedankengefüge* (1923) FREGE calls conditional sentences "*hypothetische Satzgefüge*" and what is expressed by them "*hypothetische Gedankengefüge*". He writes:[19]

> [A] hypothetical compound thought is true if its consequent is true; it is also true if its antecedent is false, regardless of whether the consequent is true or false. The consequent must always be a thought. Given [...] that "A" and "B" are sentences proper, then "not (not A and B)" expresses a hypotethical compound with the sense (thought-content) of "A" as consequent and the sense of "B" as antecedent. We may also write instead: "if B, then A." But here, indeed, doubts may arise. It may perhaps be maintained that this does not square with linguistic usage. I reply, it must once again be emphasized that science has to be allowed its own terminology, that it cannot always bow to ordinary language. Just here I see the greatest difficulty for philosophy: the instrument it finds available for its work, namely ordinary language, is little suited to the purpose, for its formation was

governed by requirements wholly different from those of philosophy. So also logic is first obliged to fashion a usable instrument from those already to hand. And for this purpose it initially finds but little in the way of usable instruments available. [...] The thought expressed by the compound sentence "If I own a cock which has laid eggs today, then Cologne Cathedral will collapse tomorrow morning" is [...] true. Someone will perhaps say: "But here the antecedent has no inner connection at all with the consequent." In my account, however, I required no such connection, and I ask that "if B, then A" should be understood solely in terms of what I have said and expressed in the form "not (not A and B)." It must be admitted that this conception of a hypothetical compound thought will at first be thought strange. But my account is not designed to square with ordinary linguistic usage, which is generally too vague and ambiguous for the purposes of logic.

Thus, as far as the truth-conditions of conditional propositions are concerned, FREGE is, whether he knew it or not, a follower of PHILO.

In turn, PEIRCE is overtly a follower of PHILO:[20]

> As far as I am concerned, I am a follower of PHILO [...]. It is completely irrelevant, whether this conception is in accordance with ordinary language.

DIODORUS OF SICILY and CHRYSIPPUS OF SOLI raised objections against PHILO's truth-functional conception of implication. While DIODORUS saw the conditional as a temporal quantification — namely by regarding a conditional true if and only if, for every instant of time t, it is not the case that at the instant t the antecedent is true and the consequent is false — CHRYSIPPUS had a virtually[21] modal conception of the conditional: according to him, a conditional is true if and only if the premise is *incompatible* with the negation of the consequence.

As we saw, FREGE and PEIRCE shared PHILO's conception. In HUGH MACCOLL's essay published in 1880 in the journal *Mind* and in his *Symbolic Logic* (1906)[22] a concept similar to that of CHRYSIPPUS can be found.

CLARENCE IRVING LEWIS, on the basis of MACCOLL's investigations, pleaded in favor of the CHRYSIPPEAN modal reading as an appropriate interpretation of the conditional, which he called "strict implication". In 1918 LEWIS published his first modal system, later called **S3**. Later on he developed his modal systems **S1** to **S5**,[23] which are stepwise based on one another. They are conceived as alternatives to the non-modal logic presented in RUSSELL's and WHITEHEAD's *Principia Mathematica* (1910-1913).[24]

LEWIS formalized strict implication in terms of negation, possibility and conjunction[25] as follows:[26]

$$p \to_s q := \neg \Diamond (p \wedge \neg q)$$

Due to the interdefinability of $\Box$ and $\Diamond$, this is logically equivalent to

$$\Box (p \to q)$$

Similarities and differences in comparison with the "material implication", as RUSSELL called the PHILONIAN conditional, are obvious:

$$p \to q := \neg (p \wedge \neg q)$$

LEWIS's comment on this was the following:[27]

> "p strictly implies q" is to mean "it is false that it is possbile that p should be true and q false" [...]
> "p materially implies q" is to mean "it is not the case that p is true and q is false."

LEWIS's main reason for formally introducing the CHRYSIPPEAN conditional were the paradoxes of material implication, in particular:

(1) $p \rightarrow (q \rightarrow p)$ (*argumentum a fortiori*),

(2) $\neg p \rightarrow (p \rightarrow q)$ (*ex falso quodlibet*).

Lewis rephrases them as follows:[28]

(1*)   If p is true, then any proposition q materially implies p.

(2*)   If p is false, then p materially implies any proposition q.

However, LEWIS admitted that the strict implication also yields the following analogous paradoxical theorems:

$$(3) \neg\Diamond\neg p \rightarrow_s (q \rightarrow p),$$
$$(4) \neg\Diamond p \rightarrow_s (p \rightarrow q),$$

in his words:[29]

(3*)   A proposition which is necessarily true is implied by any proposition [...]

(4*)   A proposition which is impossible implies any proposition.

In the later *Symbolic Logic* he will carefully discuss this issue, arguing that the paradoxicality of these theorems is just illusory: they are "paradoxical only in the sense of expressing logical truths which are easily overlooked".[30]

3 The Decision Problem and *Leibniz's Dream*

Part I of BECKER's essay pursues the goal, as we already said, of finding a propositional modal logic with a finite number of iterated modalities that is *decidable*. In the present section we focus on the *decision problem* and hint at the problems one encounters when searching for a decidable propositional modal logic *without an adequate semantic* (and *a completeness theorem*) for such a calculus being available.

The *Entscheidungsproblem* (German for "decision problem") is the question whether the first-order predicate logic is decidable. In this form the question is to be found in HILBERT & ACKERMANN *Outlines of Mathematical Logic (Grundzüge der mathematischen Logik)* published in Berlin in 1928.[31] Is there an effective method to decide the set of of logically valid formulas of first order logic?

The conceptual issues underlying the problem have a long history and may be traced back at least to LEIBNIZ. In his *Dissertatio de Arte Combinatoria* (1666)[32] early LEIBNIZ had advanced the hypothesis that all concepts could be reduced, through analysis, to a finite number of simple concepts with a procedure analogous to prime number decomposition.

Such simple concepts should have been transposed in a *lingua characteristica*, i.e. in a universal sign-system capable of directly coding simple concepts and, by means of syntactical rules, complex concepts. LEIBNIZ takes it to be possible to translate all deductive problems in the *lingua characteristica* and to decide them by means of a *calculus ratiocinator*. An example of deductive problem may be the following: "Does the conclusion C logically follow from the premises A and B? In symbols:

$$A, B \models^? C$$

A *calculus ratiocinator* may be thought of as a complex of calculation rules capable of "deciding", that is answering mechanically in the positive or in the negative, any deductive problem, i.e., applied to our schematization, whether C does follow from A and B or not. We interpret thus Leibniz's famous "*calculemus!*" Leibniz writes:[33]

> I feel that controversies can never be finished, nor silence imposed upon the sects, unless we give up complicated reasonings in favor of simple calculations, words of vague and uncertain meaning in favor of fixed symbols [*characteres*]. Thus, it will appear that every paralogism is nothing but an error of calculation [...]

If controversies were to arise, there would be no more need of disputation between two philosophers than between two calculators. For it would suffice for them to take their pencils in their hands and to sit down at the abacus, and say to each other (and if they so wish also to a friend called to help): Let us calculate without further ado!

[Sed ut redeam ad expressionem cogitationum per characteres, ita sentio nunquam temere *controversias finiri* neque *sectis* silentium imponi posse, nisi a ratiocinationibus complicatis ad *calculos* simplices, a vocabulis vagae incertaeque significationis ad *characteres* determinatos revocemur. Id scilicet efficiendum est, ut omnis *paralogismus* nihil aliud sit quam error *calculi* [...]

Quo facto quando orientur controversiae, non magis disputatione opus erit inter duos philosophos, quam inter duos Computistas. Sufficiet enim calamos in manus sumere sedereque ad abacos, et sibi mutuo (accito si placet amico) dicere: *calculemus*.]

For the sake of convenience let us depict Leibniz's idea by the following schema:

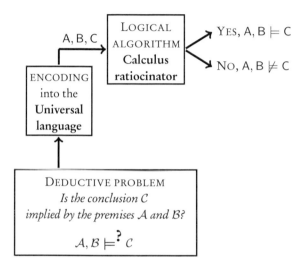

To prove the decidability of a certain set, for instance of the set of all **S3**-tautologies, it suffices to exhibit a decision algorithm for it; that is, in our example, an algorithm that takes as input a formula A in the language of **S3** and terminates its computation with a (conventionally fixed) output/answer **1/yes**, if A is a **S3**-tautology, and **0/no**, if A is not a **S3**-tautology. However, in order to prove the undecidability of a set, one must have a mathematical counterpart for the informal notion of *algorithm*

(on which the informal notion of *decidable set* depends).

OSKAR BECKER neither seems to be aware of the problem of finding an adequate formal counterpart for the concept of *decidability* nor seems to have any inkling of the necessity of making the concept of *algorithm* mathematically precise.

Thanks to the investigations of the Hilbert School, the recognition of distinct logical levels (propositional, first-order, higher-order), as it is nowadays standard, was well known since the years 1917–1919. It was clear for everyone at that time that classical propositional logic was decidable, since *there was de facto* an "effective procedure" that allowed *for any given statement* written in the language of propositional logic to decide, after a finite number of steps, whether the statement was a tautology or not, namely the *truth tables method*.

However, BECKER's problem in *On the Logic of Modalities* is to set up a propositional *modal* logic that is decidable, more precisely, to find two different extensions of **S3** that are are decidable. Such systems, which we agreed to call **S3′** and **S3″**, are, so BECKER, two *new* modal logics with 6 (**S3′**) and, respectively, 10 (**S3″**) irreducible modalities. He seems to take **S3** as undecidable on the basis of his mistaken belief that **S3** has an infinite number of iterated modalities.

Today we know that **S3** is decidable. Indeed, any (under very general conditions) propositional logic L that is finitely axiomatizable and has the finite model property is, by HARROP's theorem, decidable.[34]

However, it is not so easy to see this in the "*Survey* system", since the latter is not presented by LEWIS and, later on, by BECKER, as an extension of classical logic, as it would be natural nowadays. LEWIS and BECKER just take it to be provable that classical logic is derivable from **S3**.

Furthermore, to prove that **S3′** and **S3″** are decidable there must be an algorithm, that for each formula written in the modal language, proves whether this formula is a theorem of, respectively, **S3′** and **S3″** or not.

A complete semantic for **S3′** and **S3″** together with the finite model property would apply to obtain the decidability of theoremhood for each of these logics. But neither LEWIS provides a semantic for **S3** nor BECKER provides a semantic for **S3′** and **S3″**.

Decidability could also be proved syntactically by reducing it to terminating proof-search in an analytic (e.g. GENTZEN's style) presentation of **S3**, **S3′** and **S3″**, or by proving the equivalence between **S3**, **S3′** and **S3″** and some other logics, already known to be decidable, or by developing methods based on translations of modal logic into a fragment of first-order logic. None of these alternative was available at that time or had even been conjectured by LEWIS or BECKER.

4 Normal Modal Logics: a Quick *Resumé*

As a preliminary to the presentation and discussion, to be found in the next chapters, of BECKER's investigations on LEWIS's system **S3**, it is convenient to review here the best known *normal* modal logics, **K**, **D**, **T**, **K4**, **B**, **S4**, **S5**, as they are usually axiomatically characterized as extensions of classical logic. More precisely, the axioms comprehend all classical tautologies (or a "sufficient" selection thereof) as well as one or more additional axioms (in schematic form) that characterize the specific logic in question; the inference rules are the *modus ponens* of classical logic and one specifically modal rule, the necessitation rule

$$\frac{A}{\Box A}$$

introduced, as we already said, by GÖDEL in his 1933 paper. It says that if a proposition A is provable within the system in question, its necessitation $\Box A$ is also provable.

Once an axiom system for the minimal normal modal logic **K** is given, it is simple to give an axiomatization for the other above mentioned modal logics, for they amount to **K** with a few additional axiom schemas.

Thus, we will recall the standard axiomatization for the logic **K** as well as the modal logical schemas T, 4, B, 5 or E, and then we will see which schemas give which logic. Doing so turns out to be useful to the aim of proving, later on, to which standard modal systems BECKER's systems **S3′** and **S3″** are equivalent.

The formal modal propositional language $\mathcal{L}^\square$ is defined as usual. The alphabet contains:

- denumerably many propositional variables (or 'atoms'): $p_0, p_1, p_2 \ldots$;

- the boolean connectives: $\neg, \vee, \wedge, \rightarrow$;

- one modal operator: $\square$, for *necessity*;

- auxiliary symbols: parentheses.

The modal operator $\Diamond$ for *possibility* is conveniently not taken as primitive, and $\Diamond A$ is instead introduced as a metalinguistic abbreviation for $\neg\square\neg A$.

The set of formulas of $\mathcal{L}^\square$ is inductively defined as usual: atoms are formulas, if A and B are $\mathcal{L}^\square$-formulas then also $(\neg A), (A \vee B), (A \wedge B), (A \rightarrow B), (\square A)$ are $\mathcal{L}^\square$-formulas, and nothing else is a formula.

An axiomatic calculus for the basic axiom system **K** is set up as indicated in Table 1.

A *formal proof* in **K** is a finite list $A_1, \ldots, A_n$ of formulas such that for all i ($1 \le i \le n$): A_i is an (in-

Table 1: The calculus **K**

Axioms and axiom schemas:

— all classical tautologies

— $\Box(A \to B) \to (\Box A \to \Box B)$ (schema K)

Inference rules:

$$\frac{A \qquad A \to B}{B} \text{ MP} \qquad \text{modus ponens}$$

$$\frac{A}{\Box A} \text{ RN} \qquad\qquad \text{necessitation rule}$$

stance of) an axiom (schema) of **K**, or A_i follows by the *modus ponens* rule from two previous formulas A_j, A_k $(j, k < i)$ in the list, or A_i follows by the *necessitation rule* from a previous formula A_j $(j < i)$ in the list. A is a theorem of **K** (in symbols $\vdash_{\mathbf{K}} A$) iff there exists a formal proof $A_1, \dots, A_n$ in **K** such that A_n is A.

A modal logic which extends **K** by one ore more extra axiom schemas is called a *normal modal logic*. **K** is thus the minimal normal modal logic.

Let us now recall the modal schemas D, T, 4, B, E and their meaning:

$T : \quad \Box A \to A$

$D:$ $\quad \Box A \rightarrow \Diamond A$

$B:$ $\quad A \rightarrow \Box \Diamond A$

$4:$ $\quad \Box A \rightarrow \Box \Box A$

E $:$ $\quad \Box A \rightarrow \Box \Diamond A$

The schema T claims that *if a proposition A is necessary, then it is also true* ("ab necesse ad esse valet consequentia", in the Scholastics' reading). It is also known as "epistemic schema", since it is compatible with an *epistemic* interpretation of the modal operators. If we take "$\Box$" to be a placeholder for the operator "it is known that $(\dots)$" or "the agent x knows that $(\dots)$", the schema turns out to be in accordance with the Platonic conception of knowledge displayed in *Theaetetus* 201d- 210a, namely:

x knows *that p* if and only if

1. it is true *that p*

2. x believes *that p*

3. x is justified in believing *that p*

or, in other words, the schema T is consistent with the platonic conception of knowledge as *true belief with an account (justified)*.

The schema D claims that *if a proposition A is necessary, then it is also possible*. It is known as "deontic schema", since it is consistent with a *deontic* interpretation of the modal operators. If we take "□" to be a placeholder for the operator "it is obligatory that (...)" (and consequently "◊" to be a placeholder for the operator "it is permitted that (...)" the schema says that whatever is obligatory is also permitted.

The schema B is also known (after BECKER[35]) as *Brouwer schema*. It claims that *if a proposition A is true*, then it is *necessarily possible that it is true*.

The schemas 4 and E are both consistent with an *epistemic* interpretation of modalities and are also known as "*positive*" and "*negative*" *introspection* principle, respectively. Under an epistemic reading the schema 4 says that "if the agent x *knows that p, he also knows that he knows that p*"; while the schema E, in the equivalent reformulation:

$$\mathbf{E}' \quad : \quad \neg\Box A \to \Box\neg\Box A$$

says that *if the agent x does not know that A, then he knows that he does not know that A*.

Let us now recall the most interesting axiomatic extensions of **K** which the above schemas give rise to:

- $\mathbf{D} := \mathbf{K} + D$

- $\mathbf{T} := \mathbf{K} + T$

- $\mathbf{K4} := \mathbf{K} + 4$

- $\mathbf{B} := \mathbf{K} + T + B$

- $\mathbf{S4} := \mathbf{K} + T + 4$

- $\mathbf{S5} := \mathbf{K} + T + E$

All the systems $\mathbf{K}, \mathbf{D}, \mathbf{T}, \mathbf{K4}, \mathbf{B}, \mathbf{S4}, \mathbf{S5}$ are therefore *normal* modal logics.

The "strength" relations between them can be summarized in the following diagram, where:

- an arrow leading from a system $\mathbf{L_1}$ to a system $\mathbf{L_2}$ means that $\mathbf{L_1} \subset \mathbf{L_2}$ (that is, any $\mathbf{L_1}$-theorem is also a $\mathbf{L_2}$-theorem, but not conversely),

- the *absence* of any arrow between two systems $\mathbf{L_1}$ and $\mathbf{L_2}$ means that they are **incomparable,** that is: there is at least one $\mathbf{L_1}$-theorem which is not a $\mathbf{L_2}$-theorem, and there is as well at least one $\mathbf{L_2}$-theorem which is not a $\mathbf{L_1}$-theorem.

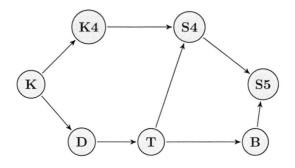

5 LEWIS's S3 and BECKER's Extensions

As we already said, the formal language and the style of axiomatization employed by LEWIS in his *Survey* and by BECKER in his *On the Logic of Modalities* are different from the now current ones.

As primitive logical operators, they take:

- the unary operators "−" and "∼", respectively for *negation* and *impossibility*,
- and the binary operators "×" and "=", respectively for *conjunction* and *(strict) equivalence*.

Thus "$-A$", "$\sim A$", "$A \times B$", "$A = B$" correspond, respectively, to "$\neg A$", "$\neg \Diamond A$", "$A \wedge B$", "$\Box(A \leftrightarrow B)$" in our notation.

The other logical boolean and "strict" operators, in particular "⊃" (*material implication*), ">" (*strict implication*) and "+" (disjunction), corresponding respectively to our "→", "→$_s$" and "∨", are instead *defined* in the expected way: $A \supset B =_{df} -(A \times -B)$, $A > B =_{df} \sim (A \times -B)$ and $A \vee B =_{df} -(-A \times -B)$.

In turn, LEWIS's (and BECKER's) axiomatization of the system **S3** is *not* given as an extension of an axiomatic calculus for classical logic by means of additional axioms and inference rules (as the axiomatizations of the normal systems reviewed in the previous chapter are). Actually, it is not at all trivial to prove that all classical tautologies are theorems of the original axiomatization of **S3**.[36]

All in all, for a contemporary reader with a basic knowledge of logic it would be cumbersome to decipher and translate in the current formalism LEWIS's and BECKER's formulas and investigations.

In this chapter, we therefore present first of all the (emended) system **S3** *not* in its original formulation, but in the *equivalent* standard formulation now current in modal logic.

Next, we shall present — again not in the original but in the now standard formulation — the axiomatization of BECKER's **S3′** and **S3″**. BECKER's claims and accomplishments will be evaluated in the next chapter.

All proofs of our assertions (e.g. as to the equivalence or non-equivalence of the logics in question) in this and the next chapter are omitted: we try to explain what is going on in *On the Logic of Modalities* in a way accessible to a wider audience.

5.1 LEWIS's *Survey System* S3

An axiomatic calculus for **S3** in the style of the most known modal logics, see Chapter 3 above, is found in Table 2.

Table 2: The calculus **S3**

Axioms and axiom schemas:

— all classical tautologies

— $\Box A \to A$ (schema T)

— $\Box(A \to B) \to \Box(\Box A \to \Box B)$ (schema K^+)

Inference rules:

$$\frac{A \qquad A \to B}{B} \text{ MP}$$

modus ponens

$$\frac{A}{\Box A} \text{ RN}^-$$

provided A is a tautology or an instance of T or K^+

Thus, **S3** contains the classical logic, the schema T, the schema K^+, the *modus ponens* and *only a restricted* version RN^- of the necessitation rule RN.

The schema K^+ is, like the schema K, a distributivity law (of the operator $\Box$ on the connective $\to$). At a a variance with K it contains a further $\Box$ after the principal connective $\rightarrow$.[37]

The inference rule RN^- says that if a proposition A is a classical tautology, or an instance of T or K^+, then its necessitation $\Box A$ is a **S3**-theorem. Thus, if "$\top$" denotes any tautology, say $p \rightarrow p$, $\Box\top$ is provable in **S3**. By contrast, $\Box\Box\top$ cannot be obtained from $\Box\top$ by means of RN^-, since it is *neither* a classical tautology nor an instance of T or K^+. Indeed, it is possible to prove that $\Box\Box\top$ is not a theorem of **S3**, which implies that the *unrestricted* necessitation rule RN is not admissible in LEWIS's system: **S3** *is not a normal modal system*.

As we said in the Introduction, the original 1918 version of **S3** contained an additional axiom, which was responsible for the modal collapse (as POST pointed out to LEWIS) and was therefore dropped by LEWIS in his 1920 "*Emendation*". In our notation, this axiom schema amounts to:[38]

$$(*) \quad (\neg\Diamond B \rightarrow_s \neg\Diamond A) \rightarrow_s (A \rightarrow_s B)$$

It is interesting to notice here how BECKER accounts for the implausibility of $(*)$, by providing the following informal, yet convincing and intriguing countermodel:[39]

> *Example*: One may think of a sequence of numbers, which is built up by drawing arbitrarily many times, one after another, num-

bered balls from an urn, whereby each time the drawn ball shall be put back in the urn before drawing the next ball. It is *unknown* how many balls are in the urn and how they are numbered. [...]

[T]he converse

$$(\sim q <\sim p) < (p < q)$$

does not hold, as our example [...] does show.

Namely, let now q and p stand for:

q : "19 appears among the first 100 places."

p : "19 appears among the first 200 places."

Then

$$\sim q <\sim p$$

does hold.

Actually, the *impossibility* that 19 occurs in the sequence (at a variance with its contingent, i.e. *incidental* not-occurring) can be due exclusively to the fact that no balls with the number 19 are contained in the urn. This impossibility holds for all places of the sequence, if it holds for some.

However, from $\sim q <\sim p$ it does *in no way* follow $p < q$, i.e. the implication: "If 19 appears among the first 200 places, then it necessarily appears within the first 100 places", since this sentence is trivially false.

5.2 BECKER's *six modalities System* S3′

Let us now consider BECKER's system **S3′**:[40]

> Before considering the real meaning of the problem of reducing the infinitely many nested modalities that arise from the iteration and composition of the symbols "∼" and "−", we shall present a purely formal investigation by which LEWIS's system becomes a closed one thanks to the addition of one further axiom. This can be done in several ways. [...] The assumptions introduced by LEWIS are (apparently) not sufficient to obtain a closed system of irreducible modalities.
>
> Therefore we add to LEWIS's axioms the new axiom 1.9:

$$-(\sim p) < \sim (\sim p)$$

BECKER is saying that — as summarized in Table 3 — this system (once formulated in our notation) is obtained from **S3** by adding one single axiom schema, namely $\Box(\Diamond A \to \Box\Diamond A)$, which is a "boxed-version" of the schema E.

Table 3: The calculus **S3′**

Axioms and axiom schemas:

— all classical tautologies

— $\Box A \to A$ (schema T)

— $\Box(A \to B) \to \Box(\Box A \to \Box B)$ (schema K^+)

— $\Box(\Diamond A \to \Box \Diamond A)$ (schema $\Box E$)

Inference rules:

$$\frac{A \qquad A \to B}{B} \text{ MP}$$ modus ponens

$$\frac{A}{\Box A} \text{ RN}^-$$ provided A is a tautology or an instance of T or K^+

BECKER gives a detailed proof of the fact that this system has 6 irreducible modalities:

- Positive modalities: $\Box A$, $\Diamond A$, A ("factual" truth);

- Negative modalities: $\neg\Box A$, $\neg\Diamond A$, $\neg A$ ("factual" falsity).

and that they are linearly ordered, as to logical strength, as follows

- Positive modalities: $\Box A \to_s A \to_s \Diamond A$
- Negative modalities: $\neg \Diamond A \to_s \neg A \to_s \neg \Box A$

5.3 BECKER's *ten modalities System* S3''

In BROUWER's and HEYTING's[41] Intuitionistic logic the *double negation principle*, $\neg\neg A \leftrightarrow A$, is not valid. More precisely, the left-to-right direction of the biconditional, $\neg\neg A \to A$, is not intuitionistically acceptable. The other direction of the biconditional,

$$A \to \neg\neg A \qquad \text{(WDN)}$$

is instead intuitionistically valid.

As we know, BECKER is also trying to explore the connection between intuitionistic and modal logic. He is thus naturally led, in particular, to interpret the intuitionistic negation ("$\neg$") — which is *stronger* than classical negation — in modal terms, as *impossibility* ("$\sim$") or, as he uses to say, *absurdity* (*Absurdität*). By replacing, in the intuitionistic law (WDN), "$\neg$" with "$\sim$" and "$\to$" with "$>$" one gets

$$A > \sim\sim A$$
(that is $\Box(A \to \neg\Diamond\neg\Diamond A)$, in our notation)

"Truth — as he puts it[42] — implies the absurdity of the absurdity (but not conversely!)".

Such principle is easily seen to be equivalent to the "boxed-version" of the schema B, $A \to \Box \Diamond A$, considered in the previous chapter. It should be now clear why BECKER called it "BROUWER's *Axiom*", a name still current in the literature.

According to BECKER, this is a reasonable axiom to consider in order to extend LEWIS's **S3**:[43]

> One can now add (this is the weakest additional postulation we propose) "BROUWER's *Axiom*" to this setting
>
> $$p = - - p < \sim\sim p \qquad (1.91)$$
>
> [...] As an [additional] axiom we choose [...]:
>
> $$\sim -p < \sim' - \sim' -p \qquad (1.92)$$
>
> [...] If one postulates $(1.91) \times (1.92)$ one can thus set up a ten modalities calculus.

In the current standard form we adopted, BECKER's second extension of LEWIS's **S3** — summarized in Table 4 — results from the latter by adding *two* axiom schemas, namely $\Box(A \to \Box \Diamond A)$, the "boxed-version" of the BROUWER's schema B, and $\Box(\Box A \to \Box \Box A)$ which is a "boxed-version" of the schema 4.

Axioms and axiom schemas:

— all classical tautologies

— $\Box A \to A$	(schema T)
— $\Box(A \to B) \to \Box(\Box A \to \Box B)$	(schema K^+)
— $\Box(A \to \Box \Diamond A)$	(schema $\Box B$)
— $\Box(\Box A \to \Box \Box A)$	(schema $\Box 4$)

Inference rules:

$$\frac{A \qquad A \to B}{B} \text{ MP}$$ modus ponens

$$\frac{A}{\Box A} \text{ RN}^-$$ provided A is a tautology or an instance of T or K^+

BECKER's claim, supported by a detailed (putative) proof, is that this system has 10 irreducible modalities:

- Positive modalities: $\Box A$, $\Diamond \Box A$, $\Diamond A$, $\Box \Diamond A$, A ("factual" truth);

- Negative modalities: $\neg\Box A$, $\neg\Diamond\Box A$, $\neg\Diamond A$, $\neg\Box\Diamond A$, $\neg A$ ("factual" falsity).

and that they are linearly ordered, as to logical strength, as follows

- Positive modalities: $\Box A \rightarrow_s \Diamond\Box A \rightarrow_s A \rightarrow_s \Box\Diamond A \rightarrow \Diamond A$

- Negative modalities: $\neg\Diamond A \rightarrow_s \neg\Box\Diamond A \rightarrow_s \neg A \rightarrow \neg\Diamond\Box A \rightarrow_s \neg\Box A$

5.4 BECKER: Further "experiments"

BECKER did also tentatively consider other two possible ways to extend LEWIS's **S3** in order to get a system with a finite number of irreducible modalities or, at least, a system with a possibly infinite number of irreducible modalities, yet all *pairwise comparable* with respect to logical strength. Here is a sketchy account of these two modal "experiments" – as he calls them.[44]

A variant of S3′

This variant, let us call it **S3′***, is proposed in a short *Observation*[45] following the presentation and the investigation of the *six modalities calculus* **S3′**. It is obtained by replacing the characteristic axiom schema $\Box E$

$$\Box(\Diamond A \rightarrow \Box\Diamond A)$$

of **S3′** (see Table 2) with the axiom schema $\Box E^*$

$$\Box(\Box\Diamond A \to \Box A)$$

BECKER's claim is that also this new system **S3′*** has 6 irreducible modalities, exactly the same as **S3′**, and that they are ordered with respect to logical strength in the same way as they are ordered in **S3′**. The only remarkable difference between the two systems, according to BECKER, is that while in **S3′** A is stronger than $\Box\Diamond A$ ($\sim\sim A$ in his notation), in **S3′*** the other way around is the case: $\Box\Diamond A$ is stronger than A. His "formalist" conclusion is:[46]

> Thus, $\sim\sim p$ is stronger than p (in contrast with BROUWER's conception). From a *purely formal point of view* it seems that also this approach can be carried through without contradiction; although it has perhaps no concrete meaning.

A more abstract approach

In §5 of Part I,[47] entitled "On the Calculus of Modalities with least Requirements, which still yields a Linear Rank Order", BECKER tentatively develops a very interesting, more abstract approach to the problem of "completing" LEWIS's calculus in such a way that in the resulting system, independently from the number (finite or infinite) of irreducible modalities, any two modalities (combinations of primitive modalities) be comparable with respect to logical strength:[48]

> One can try to free oneself from all requirements, which are imposed by special *factual* assumptions, and to seek to establish only the most general *formal* conditions of a calculus of modalities. Firstly, one can drop the demand that a reduction to finitely many fundamental modalities be possible. [...]

> On the other hand, one will have to keep the demand of a linear rank order of the modalities, whereby the implicational relation of any two distinct modalities is uniquely determined. Otherwise a modality cannot be uniquely determined by its "rank of logical strength" anymore. This claim should be in any case the upper bound of our formal freedom.

> Now, on one hand the question is whether the LEWISIAN Calculus satisfies this demand **and, on the other hand, whether the natural axioms of a theory of the rank order of the modalities cannot be established independently from the LEWISIAN Calculus**. The answer to the first ques-

tion is negative, the answer to the second
question is affirmative.

BECKER conveniently uses here as basic, elementary positive modality the operator "$\Box$" ("N" in
his symbolism, corresponding to LEWIS's "$\sim$ $-$").
He then fixes a number of "rules" to which composed ("non-elementary") modalities in the calculus
of modalities should obey.

By a *composed modality* he means a finite, possibly empty string of $\Box$'s and $\neg$'s (e.g. $\Box\Box$, $\neg\Box\neg$,
$\neg\Box\neg\Box\neg\Box$, …). Composed modalities are denoted
by capital Greek letters $\Lambda, \Pi, \ldots$. Given two composed modalities Λ and Π, the composed modality
arising from their juxtaposition (in the given order)
is denoted by $\Lambda\Pi$.[49] A composed modality Λ is *positive* (*negative*) if and only if it contains an *even* (*odd*)
number of $\neg$'s. Thus e.g. $\Box$ and $\neg\Box\neg$ (that is: $\Diamond$)
are positive, while $\Box\neg$ (equivalent to $\sim$, *impossible*)
is negative.

The rules are intended to impose a number of conditions concerning the *preservation within the calculus of relations of logical strength between modalities
under composition/juxtaposition*,[50] and can be equivalently rephrased as follows.

For any basic or composed modalities Λ, Λ' and Π,

if

$$\Lambda A \rightarrow_s \Lambda' A$$

is a theorem of the calculus of modalities for *every* formula A

then also

- $\Lambda\Pi A \to_s \Lambda'\Pi A$ [rule R_1]

- $\Pi\Lambda A \to_s \Pi\Lambda' A$, where Π is *positive* [rule R_2]

- $\Pi\Lambda' A \to_s \Pi\Lambda A$, Π is *negative* [rule R_3]

shall be theorems of the calculus of modalities for *every* formula A.

In other words, these three rules say that if the modality Λ turns out to be at least as strong as the modality Λ' in the calculus, then also $\Lambda\Pi$ shall be at least as strong as $\Lambda'\Pi$ (R_1), and $\Pi\Lambda$ shall be at least as strong as $\Pi\Lambda'$ when Π is positive (R_2), as well as $\Pi\Lambda'$ shall be at least as strong as $\Pi\Lambda$ when Π is negative (R_3).

Now, the point is that while LEWIS's calculus is closed under these three rules, it contains *incomparable* modalities. So, this is BECKER's very interesting idea, one should try to devise the "weakest possible axiomatic conditions" one should add to the above rules $R_1 - R_3$, *in order to obtain a calculus in which all the modalities are linearly ordered*, that is are pairwise comparable in strength.

At the end of a rather elaborate argument, he arrives at the claim that a stepwise generalization of BROUWER's schema (in the form $\Box B$, see above):

$B_1 = \Box(A \to \Box \Diamond A)$ (that is $\Box B$)

$B_2 = \Box(A \to \Box \Box \Diamond A)$

$B_3 = \Box(A \to \Box \Box \Box \Diamond A)$

$\vdots$

provides an infinite number of axiomatic conditions which, added to LEWIS's **S3** together with the above rules (R_1)–(R_3), produce a calculus — let us call it **SM**, for further reference — whose modalities, although infinite in number, are linearly ordered. He leaves as an open problem the question whether all these infinite modalities of **SM** are irreducible.

6 BECKER's accomplishments: An Assessment

The evaluation of BECKER's formal investigations, of the results and claims we illustrated in the previous chapter, may well start from the questions raised by GÖDEL in his *Review* of *On the Logic of Modalities*:[51]

> [T]he author proposes various additional axioms and then seeks to specify a system, with as few assumptions as possible, for which a linear ordering [of modalities with respect to logical strength] still exists. All in all, three different kinds of the calculus of modalities emerge [...]. As far as the purely formal side is concerned, one can hardly take exception to anything here, but there remain essential gaps to be filled in, some of which the author himself points out. Above all, it is nowhere shown that the three systems set up really differ from one another and from Lewis's system (in other words, that the additional axioms are not in fact equivalent and do not follow from Lewis's); nor, furthermore, that the six, or ten, basic modalities obtained cannot be still further reduced.

As we have seen, BECKER actually proposed *four* calculi altogether: **S3′** (the *six modalities calculus*, see 5.2), **S3″** (the *ten modalities calculus*, see 5.3), the variant **S3′*** of **S3′** (see 5.4) and **SM** (the calculus with infinitely many linearly ordered modalities, see 5.4). It is likely that "the three different kinds of the calculus of modalities" GÖDEL is hinting at are the extensions **S3′**, **S3″** and **SM** of **S3**. Anyway, including also **S3′***, the questions are:

(i) Are **S3′**, **S3″**, **S3′*** and **SM** pairwise non-equivalent?

(ii) Can the additional axioms of these systems be derived from **S3**?

(iii) Are the claimed "irreducible" modalities of these calculi really irreducible?

The deciding answers to these questions follow from the following three results:

1. Both S3′ and S3″ are equivalent to the system S5. S3′ is indeed *one of the two equivalent axiomatizations of the normal modal system* **S5** (see Chapter 4), as it is explicitly introduced for the first time (and thus named) in 1932 in the *Appendix II* of LEWIS & LANGFORD *Symbolic Logic*.[52] *The other equivalent axiomatization of* **S5** *indicated there is exactly* BECKER*'s system* **S3″**! So, the *six* and the *ten modalities* calculi proposed by BECKER are in fact different axiomatizations of one and the same modal system, and while the six "irreducible" modalities of **S3′** are really irreducible, the ten "irreducible" modalities of **S3″** of course boil down to the six of **S3′**, alias **S5**. This said, and given the fact that BECKER's axiom schemas $\Box B$, $\Box 4$ and $\Box E$ are indeed *not derivable* from **S3**,[53] we have a conclusive answer to questions (i)–(iii) above as to **S3′** and **S3″**.

2. S3′* collapses. This was first observed by PARRY in 1939, while one year earlier CHURCHMAN[54] was still conjecturing that **S3′*** had an infinite number of irreducible modalities. Indeed, using the **S3**-theorem $\Box\Diamond(A \to \Box A)$, PARRY was able to show that $A \to_s \Box A$ is a theorem of **S3′*** and thus "reduces to the system of material implication".[55] Of course, this also shows (see questions (ii) and (iii) above) that the additional axiom $\Box E^*$ of **S3′*** *cannot* be derived from **S3**, and that the six "irreducible" (according to BECKER) modalities of **S3′*** boil down to two, that is (actual) truth and falsity.

3. SM is equivalent to the system S5. Finally, as far as the fourth "system of modalities", **SM**, tentatively proposed by BECKER, CHURCHMAN proved in 1938[56] that it is equivalent to **S3′** — that is, by what we said above, to **S5**. *This claim is indeed correct*, although CHURCHMAN's proof thereof is not, because he did not adequately formalize the system **SM** as BECKER intended it.[57]

All in all, we can say that BECKER's claims about **S3′** *were right* — and so also that *he was indeed the first*, two years before the official birthdate of the system **S5**, to identify this modal system and to investigate some of its properties.

BECKER's claims about **S3″**, **S3′*** and **SM** *were instead wrong*. Yet, concerning **S3″** the proofs he gave to support his claims were correct, except that he did not notice that the characteristic schema $\Box E$ of **S3′** can be (easily) derived from the two characteristic schemas $\Box B$ and $\Box 4$ of **S3″**, modulo LEWIS's system **S3**. Concerning **SM**, again, his proof that the (supposedly infinite) modalities of this system are linearly ordered is very clever, and correct. Unfortunately he did not notice that, on the basis of **S3**, already the first two schemas, B_1 and B_2, of the infinite sequence of schemas $\{B_n\}_{n \geq 0}$ he postulated, together with the rule R_2 are sufficient to prove the schema $\Box 4$ — and so to make **SM** equivalent to **S3′**, alias **S5**. Whether the same does happen with a basis *weaker* than **S3** is an interesting question, open as far as we know and worth to be investigated.

In the light of these results — of the four systems he proposed, three are in fact equivalent while the remaining one collapses —, one might be tempted to underrate BECKER's *formal* contributions in *On the Logic of Modalities*. On the contrary, and notwithstanding these shortcomings, BECKER's *pioneering* work, containing sophisticated insights and interesting technical solutions, has played an extremely important role in the *early* development of modal logic in the decade 1930–1940, as witnessed by the scientific contributions of other scholars who, at that time, referred to BECKER's investigations and to the problems raised by him, and took them as a basis for further developments and investigations.[58]

Last but not least, BECKER was the first to advance the idea of a *modal interpretation of intuitionistic logic* — the reader should keep in mind that BECKER was writing in 1930, and that the birthdate of intuitionistic logic as a formalized system is 1928, thanks to HEYTING's axiomatization.[59]

The idea is exposed and elaborated in the *Appendix to the Part I*[60] of *On the Logic of Modalities*:[61]

> How is now the HEYTINGIAN calculus related to the uncompleted and the completed LEWISIAN calculus?
>
> Firstly, the question of an appropriate "translation" of the symbols emerges.

More precisely, his idea is to define a vocabulary translation associating to each intuitionistic logical operator ($\rightarrow$, $\vee$, $\wedge$, $\neg$) a corresponding logical operator of LEWIS's calculus **S3**, in such a way that every theorem of the intuitionistic calculus **H** becomes, once transformed according to the translation, a theorem of **S3**. BECKER tentatively considers three candidate translations:[62]

(T1) H: $\to$, $\vee$, $\wedge$, $\neg$ $\Rightarrow$ L: $\to$, $\vee$, $\wedge$, $\neg$

(T2) H: $\to$, $\vee$, $\wedge$, $\neg$ $\Rightarrow$ L: $\to_s$, $\vee_s$, $\wedge$, $\neg\Diamond$ (where $A \vee_s B =_{df} \Box(A \vee B)$)

(T3) H: $\to$, $\vee$, $\wedge$, $\neg$ $\Rightarrow$ L: $\to$, $\vee$, $\wedge$, $\neg\Diamond$

As to (T1), he observes that the T1-translation of every intuitionistic theorem is *obviously* a **S3**-theorem, because intuitionistic logic is included in classical non modal logic, and the latter in turn is included in the LEWIS's system. On the other side, he rightly stresses that[63]

> [...] this is a worthless triviality. Indeed, the purpose of a comparison between intuitionistic and modal logic can only be to make the deficits of the former with respect to the latter comprehensible by interpreting the intuitionistic notions by the specific modal-logical notions, that is the LEWISIAN "strict" notions (strict implication, strict logical sum, impossibility).

Concerning the second translation, he again rightly observes that the T2-translation $A \vee_s A \leftrightarrow_s A$ of $A \vee A \leftrightarrow A$, which is one of the axioms of HEYTING's calculus, *is not a theorem* of **S3** — the latter would collapse otherwise!

Finally, as far as the third of the proposed translations is concerned, BECKER claims (without giving a detailed proof) that the T3-translation of the axiom $(A \rightarrow B) \wedge (A \rightarrow \neg B) \rightarrow \neg A$, that is the formula

$$(*) \qquad (A \rightarrow B) \wedge (A \rightarrow \neg \Diamond B) \rightarrow \neg \Diamond A$$

is not a theorem of **S3**, and thereby concludes his "translation-experiments" as follows:[64]

> At this point a further investigation must begin, with the aim to assess *whether and which additions must be made to the extended* LEWISIAN *System (Calculus of 10 Modalities, Calculus of 6 Modalities) so that the* HEYTINGIAN *Axiom (11)* [i.e. (∗) above] *holds.* Further problems can nevertheless arise because of the difference of the undefined notions in the HEYTINGIAN and the LEWISIAN System. The solution of these tasks and the overcome of these difficulties shall be left to future work.

Indeed, it is not difficult to prove that BECKER was right in claiming that (∗) is underivable in **S3**. Actually, it is possible to prove even more: *every normal modal system containing the schema T and* (∗) *collapses.* This implies that also with respect to the extended LEWISIAN systems mentioned by

BECKER the translation T3 would boil down to the *trivial* translation T1.

To conclude, let us tell the end of the story: only three years later someone else, namely KURT GÖDEL, did the "future work" and "found the solution of these tasks" as predicted by BECKER. In the already mentioned *An interpretation of the intuitionistic propositional calculus*[65] GÖDEL, who is his 1931 *Review* of *On the Logic of Modalities* had hastily mentioned and too roughly dismissed BECKER's idea of a *modal interpretation of intuitionistic logic*, saying[66]

> [...] the author discusses, from a formal as well as a phenomenological standpoint, the connections that in his opinion obtain between modal logic and the intuitionistic logic of Brouwer and Heyting. It seems doubtful, however, that the steps here taken to deal with this problem on a formal plane will lead to success,

provided the first *sound and faithful translation* of propositional intuitionistic logic into a modal system, namely **S4**.[67] In the paper, BECKER is mentioned for having introduced the axiom $\Box A \rightarrow_s \Box\Box A$ ($\Box 4$, see 5.3) but, quite unfairly, *not for having anticipated the very idea of a modal translation of intuitionistic logic*. By the way, as was later proved by HACKING,[68] it is also possible to define a sound and faithful translation of intuitionistic logic even in the LEWIS's modal system **S3** — as BECKER had tried to do.

Notes

1. BECKER 1930. The *Yearbook for Philosophy and Phenomenological Research (Jahrbuch für Philosophie und phänomenologische Forschung)* was founded by EDMUND HUSSERL in 1912 and served the HUSSERL's circle as an important organ during HUSSERL's Freiburg period (1916–1938). The first issue of the journal was published in 1913 and contains HUSSERL's *Ideas for a Pure Phenomenology and Phenomenological Philosophy*. Volume 8 includes HEIDEGGER's masterpiece *Being and Time* (1927) as well as OSKAR BECKER's famous investigation on the logic and ontology of mathematical phenomena "*Mathematical Existence (Mathematische Existenz)*."

2. BECKER 1914.

3. BECKER 1923.

4. BECKER 1927. Hereto see at least: GETHMANN 2003, PECKHAUS 2005, MITTELSTRASS & GETHMANN-SIEFERT 2002.

5. BECKER 1952; cp. MARTIN 1969. For a complete bibliography of BECKER's works see ZIMNY 1969.

6. MACCOLL 1906.

7. LEWIS 1918.

8. The name appears for the first time in Appendix II of LEWIS & LANGFORD 1932.

9. LEWIS 1920.

10. Cp. CRESSWELL et al. 2016, 281 f.

11. BECKER 1930, 4.

12. GÖDEL 1931.

13. Hereto cp. GÖDEL 1931.

14. PARRY 1939.

15. GÖDEL 1933.

16. GÖDEL 1933, 301.

17. BOCHEŃSKI 1956, 116.

18. Loc. cit., 117.

19. FREGE 1923, 46.

20. PEIRCE 1992, 125 f.

21. If we assume that "incompatibility" and "impossibility" mean the same.

22. MACCOLL 1880, MACCOLL 1906.

23. LEWIS & LANGFORD 1932.

24. RUSSELL & WHITEHEAD 1910–13.

25. In order to comply with the now current logical notation, we prefer not to adopt LEWIS's symbolic apparatus. In particular, we use "$\rightarrow$", "$\rightarrow_s$", "$\neg$", "$\wedge$", "$\vee$", "$\Box$" and "$\Diamond$" to denote, respectively, material implication, strict implication, conjunction, disjunction, necessity and possibility.

26. LEWIS & LANGFORD 1932, 124.

27. Loc. cit. 124, 136.

28. Loc. cit. 142.

29. Loc. cit. 174.

30. C. I. LEWIS & C. H. LANGFORD 1932, 248 ff.

31. HILBERT & ACKERMANN 1928, 4; 7-9.

32. LEIBNIZ 1666.

33. LEIBNIZ 1688, 912–913.

34. A logic L has the finite model property if any non-theorem of L is falsified by some finite model of L.

35. The reason why BECKER uses this name for the schema is explained in the next chapter.

36. See LEWIS & LANGFORD 1932, 136 ff.

37. Notice that K is provable in **S3**, by using K^+, T and the transitivity of implication.

38. That is, $(\sim q <\sim p) < (p < q)$ in the LEWIS-BECKER symbolism. Notice that the converse of $(*)$, $(A \to_s B) \to_s (\neg \Diamond B \to_s \neg \Diamond A)$ is a theorem of **S3** (actually an axiom, $(p < q) < (\sim q <\sim p)$, in LEWIS's presentation).

39. BECKER 1930, 8-9.

40. BECKER 1930, 11-12. Recall that "$-A$" corresponds to our "$\neg A$".

41. BECKER refers explicitly to HEYTING 1930, which contains the first (complete) presentation of intuitionistic logic as a formalized calculus. The paper was published in the same year of *On the Logic of Modalities*, but was circulating since 1928.

42. BECKER 1930, 17.

43. BECKER 1930, 17-18.

44. BECKER 1930, 2.

45. BECKER 1930, 15-16.

46. BECKER 1930, 16.

47. BECKER 1930, 25-30.

48. BECKER 1930, 25.

49. E.g., for $\Lambda = \Box\Box$ and $\Pi = \neg\Box\neg$, we have $\Lambda\Pi = \Box\Box\neg\Box\neg$.

50. These rules have not been correctly interpreted and formalized in CHURCHMAN 1938, the first (and unique, as far as we know) paper where this experiment by BECKER is detailedly analyzed. Incidentally, notice that the inference rules

$$\frac{A \to_s B}{\Box A \to_s \Box B} \quad \text{and} \quad \frac{A \to_s B}{\Diamond A \to_s \Diamond B}$$

known also in the current literature as BECKER's *rules*, were given this name in CHURCHMAN 1938 (cp. HUGHES & CRESSWELL 1996, 200, 207) because (uncorrectly) regarded as specific instances of rule (R_2), see below, of BECKER.

51. GÖDEL 1931, 6.

52. LEWIS & LANGFORD 1932, 501.

53. The irreducibility of the six modalities of **S5**, alias **S3′**, and the other facts and claims we mentioned are proved in LEWIS & LANGFORD 1932, 497 ff.

54. CHURCHMAN 1938, 78.

55. PARRY 1939, 153 f. The proof is easy, using our "standard" formulation of **S3′***: **S3′*** $\vdash \Box(\Box\Diamond(A \to \Box A) \to \Box(A \to \Box A))$ since this is an instance of BECKER's schema $\Box E^*$, hence (by using the schema T) **S3′*** $\vdash \Box\Diamond(A \to \Box A) \to \Box(A \to \Box A)$ and so, by *modus ponens* with the **S3**-theorem $\Box\Diamond(A \to \Box A)$, one has **S3′*** $\vdash \Box(A \to \Box A)$, that is $A \to_s \Box A$.

56. CHURCHMAN 1938, 78 ff.

57. As we explained in Chapter 5, fn. 50. One can prove that **SM** (as an extension of LEWIS's **S3**) boils down again to **S5** as follows. By $\Diamond\Box A \to_s A$, which is **S3**-strictly equivalent to an instance of the schema B_1 of **SM**, we get by the rule R_2 ($\Box\Box$ being positive) **SM** $\vdash \Box\Box\Diamond\Box A \to_s \Box\Box A$. On the other side, **SM** $\vdash \Box A \to_s \Box\Box\Diamond\Box A$ because it is an instance of the axiom schema B_2. Thus, by transitivity of $\to_s$, **SM** $\vdash \Box A \to_s \Box\Box A$, that is: **SM** proves the schema $\Box 4$. But the latter, together with B_1 ($\Box B$), gives BECKER's **S3′** which in turn, as we have seen above, is equivalent to **S5**.

58. In this regard, FEYS 1937 and 1938 should at least be added to the already mentioned works GÖDEL 1931 and 1933, LEWIS & LANGFORD 1932, CHURCHMAN 1938, PARRY 1939 referring to, and discussing, BECKER's work.

59. Published two years later in HEYTING 1930.

60. BECKER 1930, 30-35.

61. BECKER 1930, 31.

62. Here rephrased in the non-Lewisian symbolism so far employed.

63. BECKER 1930, 31.

64. BECKER 1930, 33.

65. GÖDEL 1933.

66. GÖDEL 1931, 6.

67. GÖDEL's translation τ from the formulas of the language $\mathcal{L}$ of intuitionistic propositional logic to the formulas of the modal propositional language $\mathcal{L}^\square$ of **S4** (see Chapter 4) is such that, for every $\mathcal{L}$-formula A, the following does hold: (i) if $\vdash_{\mathbf{H}} A$ then $\vdash_{\mathbf{S4}} \tau(A)$ (*soundness*), and (ii) if $\vdash_{\mathbf{S4}} \tau(A)$ then $\vdash_{\mathbf{H}} A$ (*faithfulness*). Point (ii) was only conjectured by GÖDEL, and proved 15 years later in MCKINSEY & TARSKI 1948.

68. HACKING 1963.

Bibliography

[Becker, 1914] Becker, O. (1914). *Über die Zerlegung eines Polygons in exclusive Dreiecke auf Grund der ebenen Axiome der Verknüpfung und Anordnung.* F. A. Brockhaus, Leipzig.

[Becker, 1923] Becker, O. (1923). *Beiträge zur phänomenologischen Begründung der Geometrie und ihrer physikalischen Anwendung.* Jahrbuch für Philosophie und phänomenologische Forschung VI. Max Niemeyer Verlag, Halle.

[Becker, 1927] Becker, O. (1927). *Mathematische Existenz. Untersuchungen zur Logik und Ontologie mathematischer Phänomene.* Jahrbuch für Philosophie und phänomenologische Forschung VIII. Max Niemeyer Verlag, Halle.

[Becker, 1930] Becker, O. (1930). *Zur Logik der Modalitäten.* Jahrbuch für Philosophie und phänomenologische Forschung XI. Max Niemeyer Verlag, Halle.

[Becker, 1952] Becker, O. (1952). *Untersuchungen über den Modalkalkül.* Westkulturverlag Anton Hain, Meisenheim/Glan.

[Bocheński, 1956] Bocheński, J. M. (1956). *Formale Logik*. Verlag Karl Alber, Freiburg (Breisgau)/München. (Engl. transl.: *History of Formal Logic*, Notre Dame 1961).

[Churchman, 1938] Churchman, C. W. (1938). On finite and infinite modal systems. *The Journal of Symbolic Logic*, 3:77–82.

[Cresswell et al., 2016] Cresswell, M., Mares, E., and Rini, A., editors (2016). *Logical Modalities from Aristotle to Carnap: The Story of Necessity*. Cambridge University Press, Cambridge.

[Feferman et al., 1986] Feferman, S., Kleene, S., Moore, G., Solovay, R., and van Heijenoort, J., editors (1986). *Kurt Gödel Collected Works. I: Publications 1929–1936*. Oxford University Press, Oxford.

[Feys, 1937] Feys, R. (1937). Les logiques nouvelles des modalités. *Revue néo-scolastique de philosophie*, 56:517–553.

[Feys, 1938] Feys, R. (1938). Les logiques nouvelles des modalités (suite et fin). *Revue néo-scolastique de philosophie*, 58:217–252.

[Frege, 1923] Frege, G. (1923). Gedankengefüge. *Beiträge zur Philosophie des deutschen Idealismus*, III:36–51.

[Gethmann, 2003] Gethmann, C. F. (2003). Hermeneutic phenomenology and logical intuitionism: On Oskar Becker's Mathematical Existence. *New Yearbook for Phenomenology and Phenomenological Philosophy*, 3:143–160.

[Gödel, 1931] Gödel, K. (1931). Besprechung von *Becker 1930*: Zur Logik der Modalitäten. *Monatshefte für Mathematik und Physik (Literaturberichte*, 38:5–6. (Reprinted in Feferman et. al. 1986, 216–217).

[Gödel, 1933] Gödel, K. (1933). Eine Interpretation des intuitionistischen Aussagenkalküls. *Ergebnisse eines mathematischen Kolloquiums*, 4:39–40. (Reprinted in Feferman et. al. 1986, 300–301).

[Hacking, 1963] Hacking, I. (1963). What is strict implication? *The Journal of Symbolic Logic*, 28:51–71.

[Heyting, 1930] Heyting, A. (1930). Die formalen regeln der intuitionistischen logik. *Sitzungsberichte der Preussischen Akademie der Wissenschaften. Physikalisch-mathematische Klasse*, pages 42–56, 57–71, 158–169.

[Hilbert and Ackermann, 1928] Hilbert, D. and Ackermann, W. (1928). *Grundzüge der theoretischen Logik*. Springer, Berlin.

[Hughes and Cresswell, 1996] Hughes, G. E. and Cresswell, M. J. (1996). *A New Introduction to Modal Logic*. Routledge, London and New York.

[Leibniz, 1666] Leibniz, G. W. (1666). *Dissertatio de Arte Combinatoria.* J. S. Fickium et J. P. Seuboldum, Lipsiae. (Akad. A VI 1, 163–230).

[Leibniz, 1688] Leibniz, G. W. (1688). *De arte characteristica ad perficiendas scientias ratione nitentes.* Akad. A VI 4 nr. 912.

[Lewis, 1918] Lewis, C. I. (1918). *A Survey of Symbolic Logic.* University of California Press, Berkeley.

[Lewis, 1920] Lewis, C. I. (1920). Strict implication – an emendation. *Journal of Philosophy, Psychology and Scientific Methods*, 17:300–302.

[Lewis and Langford, 1932] Lewis, C. I. and Langford, C. H. (1932). *Symbolic Logic.* The Century Co., New York. (Dover Publications, New York ^{2}1959).

[MacColl, 1880] MacColl, H. (1880). Symbolic reasoning, I. *Mind*, 5:45–60.

[MacColl, 1897] MacColl, H. (1897). Symbolic reasoning, II. *Mind*, 6:493–510.

[MacColl, 1900] MacColl, H. (1900). Symbolic reasoning, III. *Mind*, 9:75–84.

[MacColl, 1906] MacColl, H. (1906). *Symbolic Logic and Its Applications.* Longmans, Green & Co., New York.

[Martin, 1969] Martin, G. (1969). Oskar Beckers Untersuchungen über den Modalkalkül. *Kant-Studien*, 60:312–318.

[McKinsey and Tarski, 1948] McKinsey, J. C. C. and Tarski, A. (1948). Some theorems about the sentential calculi of Lewis and Heyting. *The Journal of Symbolic Logic*, 13:1–15.

[Mittelstraß and Gethmann-Siefert, 2002] Mittelstraß, J. and Gethmann-Siefert, A., editors (2002). *Die Philosophie und die Wissenschaften. Zum Werk Oskar Beckers*. Wilhelm Fink Verlag, München.

[Parry, 1932] Parry, W. T. (1932). Zum Lewisschen Aussagenkalkül. *Ergebnisse eines mathematischen Kolloquiums*, 5:15–16.

[Parry, 1939] Parry, W. T. (1939). Modalities in the survey system of strict implication. *The Journal of Symbolic Logic*, 4:137–154.

[Peckhaus, 2005] Peckhaus, V., editor (2005). *Oskar Becker und die Philosophie der Mathematik*. Wilhelm Fink Verlag, München.

[Peirce, 1992] Peirce, C. S. (1992). *Reasoning and the Logic of Things. (The Cambridge Conferences Lectures of 1898)*. Harvard University Press, Cambridge (Massachusetts)/London.

[Whitehead and Russell, 1910-13]
Whitehead, A. N. and Russell, B. (1910-13). *Principia Mathematica.* Cambridge University Press, London.

[Zimny, 1969] Zimny, L. (1969). Oskar Becker – Bibliographie. *Kant-Studien*, 60:319–330.

Bandaufstellung

Alle erschienenen Bücher können unter der angegebenen
ISBN direkt online (http://www.logos-verlag.de) oder
per Fax (030 - 42 85 10 92) beim Logos Verlag Berlin
bestellt werden.